On Interaction

Greg Feild

April 21, 2018

The history of natural philosophy is characterized by the interplay of two rival philosophies of time - one aiming at its "elimination" and the other based on the belief that it is fundamental and irreducible. _____

The basic objection to attempts to deduce the unidirectional nature of time from concepts such as entropy is that they are attempts to reduce a more fundamental concept to a less fundamental one.

-- G. J. Whitrow

Time is invention or it is nothing at all. But of time-invention physics can take no account … *Modern physics … rests altogether on a substitution of time-length for time invention.*

-- Henri Bergson

Apart from time there is no meaning for purpose, hope, fear, energy. If there be no historic process, then everything is what it is, namely, a mere fact. Life and motion are lost.

-- Alfred North Whitehead

Science in its effort to become more "rational" tends more and more to suppress variation in time.

-- Emile Meyerson

It cannot be too often emphasized that physics is concerned with the measurement of time, rather than with the essentially metaphysical question as to its nature … We must not believe that physical theories can ultimately solve the metaphysical problems that time raises.

-- Mary F. Cleugh

Source (cherry picked from) : *Physics and the Ultimate Significance of Time*
Edited by David R. Griffin
State University of New York Press; 1986

Abstract:

This book examines and explores the
"Universal Model of Our Sinister Universe".

We will discuss the nature of our new model,
the general nature of physical models,
the nature of physics, the nature of science,
and consequently, and unavoidably, human nature.

:)

In order to seek truth it is necessary once in the course
of our life to doubt as far as possible all things.

-- Rene Descartes

About the author:

Greg Feild writes for posterity,   : (
but he hopes, and expects,
to find readers today!

He also designs his own book covers.

Coming soon:

"On Rotation"

It was a dark, and stormy night, in the sinister universe. Our hero, our *bulldog*,
clutched their satchel ever closer; as inside was,

the theory of everything.

Our *tenacious* hero, held, not *just,* the theory of everything, they carried

the *quantum* theory of everything.

Thunder crashed like colliding stars . . .

Introduction:

Old habits die hard.

Old systems of belief, and old patterns of thought, seem to die even harder!

Everyone is familiar with the story of the phlogiston.

Everyone knows the story of the phlogiston, because everyone tells the story of the phlogiston to illustrate the historical errors in people's interpretation of physical process, and the subsequent 'paradigm shifts' in physics.

The phlogiston, the crystal spheres, the electromagnetic ether, … ; to this growing list of missteps and misunderstandings (i.e. *misinterpretations*), we must add fields, the fabric of spacetime, and the phantasmagorical wave function.

Our new <u>physical model</u> of the world, is not so much a paradigm shift, as a purging of all the current psychedelic, pseudo-scientific silliness sullying science;

An exorcism of the painfully embarrassing, paranormal phenomena, plaguing (and continually accruing to!) physics since the infamous Copenhagen conference.

We are *determined* to return *everyone* to the days of 'old school' physics, where, at the very least, physical theories dealt with physical phenomena, and scientific theories were confronted with experimental data.

So, what is the phlogiston? No one remembers …

Phlogiston.   It's a funny word!

Let's do physics!

:)

The slippery slope:

When *did* physics begin the awful slide into inappropriate, unschooled speculation, and unfettered, unfounded, and unwarranted (and unbelievable!) supposition?

We believe, it all began the day physicists "boldly and triumphantly" dismissed the electromagnetic ether, and decided to let the electromagnetic fields "stand on their own".

Unfortunately, on that fateful day, they threw away the baby, but kept the bathwater !

(No actual babies were harmed in the employment of this metaphor.)

"We *now* know", the wave like appearance and properties of electromagnetic radiation can be accounted for by;

1) the periodic emission of photons from an oscillating source, and
2) the periodic oscillation of the polarization of the emitted photons.

Both these oscillations are of the same frequency; nu = E/h

It seems odd, on the discovery of the photon, that people did not immediately revisit the electromagnetic field theory.

Perhaps they had forgotten their initial discomfort when creating real electromagnetic fields existing independently in time and space.

Perhaps they were distracted, their incredulity dulled, by collapsing wave functions, ghostly, part-time particles, and the advent of the 'fabric' of time and space.

It must have been a heady, exciting, and *confusing* time, to be sure !

The result: "modern physics" is an *incoherent*, hot mess.

… and still confusing.

Just saying.

Symmetry:

The fundamental symmetries of the universal model, are the symmetries of space and time.

Space is flat, isotropic, homogeneous, smooth, continuous; Euclidean. Space is eternal and unchanging; from the past, through the present, and into the future. The space of today is the space of ten bazillion years ago.   Space will be the same tomorrow.

Space is **where things happen**.   TIme is **things happening**.

In *any* inertial reference frame (and we always choose the *universal* inertial reference frame), time is smooth, continuous, isotropic, homogeneous; Newtonian. Time is eternal, smooth, and unchanging; from the past, through the present, and into the future. The rate of flow of time of yesterday is the same as the rate of flow of time of ten bazillion years ago. Time will *continue* to flow in exactly the same way tomorrow.

These facts about time and space are what make the the derivation of the differential calculus, *and* the ability of the differential calculus to describe the rates of change of real world physical processes and dynamical systems, possible.

Now is the time for plain, pedantic, didactic, discourse. And *italics*, and **boldface type** !

The time of physics is (or should be) the quantification of ordinary time that people use to keep appointments, measure heart rates, the speed of race cars and sprinters, how long one can hold their breath, etc.

Time is *manifest* as a sequence of events; as a succession of moments.

Time *is* the ticking clock.

The time reversal invariance of our physical equations, represents the fact that the flow of time is a smooth, continuous, invariant, invariable, un-variable, relentless, inevitable, enduring, unchanging, and *constant* process.

This notion of time is the foundational basis of the calculus. Without this notion of time we *cannot* do physics.

If the equations of physics were *not* invariant under time reversal, this would imply that these equations were not employing our common, intuitive, and ordinary notions of time *correctly*.

Time is symmetric, from the past, through the present, and into the future, because it is a *constant* 'flow'. This fact *alone* allows us to cast our mathematical physical equations in terms of this important variable.

Time flows forward, **marking change**. *We are <u>not</u> allowed to <u>assume</u>* that time could/would/should run backwards, just because our mathematical equations seem to "allow" it.

This would be an incorrect *interpretation* of the math.

Time flows forward due to the conservation of energy and <u>*momentum*</u>.

It is really as simple as that.   Really.

(The proverbial smashed vase will <u>never</u> reassemble itself.)

As for space:

One can only answer the questions; "How much space is over there?",
                                        "How much space do you *see*?",

with the questions, "Between what and what and what?" "Bounded by what?"

One can measure a particular space by measuring the time taken for an object of known, fixed velocity (i.e. the photon), to travel from point A to point B, etc.

The fact that different observers, in different inertial reference frames, measure different time intervals for the same 'event', is analogous to the fact that objects weigh less on the moon than they do on the earth.

In the first instance, the nature, flow, and experience of Newtonian time is exactly the same in both reference frames, even though the observers may measure different relative times (*and positions*) for any given 'event'.

In the second instance, the mass of the object is constant and the relative difference in the measured weights is dependent on the choice of environment.

This is why, in the universal model, our absolute inertial reference frame is always taken to be the fixed background of space.

This is where we must put our fixed Cartesian coordinate systems; and *not* fixed at the center of a rotating body, and *never* at the center of a *massive* rotating body, in order to do physics correctly, and obtain the most accurate results possible from our calculations.

The second symmetry of the universal model is -

For every action there is an equal and opposite reaction.

This is true for the usual, central, radial force between two interacting bodies, *and* for the velocity dependent forces arising from the influence of the magnetic field vectors.

We know the magnetic forces "do no work", hence any angular motion induced in a first body due to the magnetic field of a second body, must be compensated for by an equal and opposite change in the angular motion of the second body due to the magnetic field of the first body.

Think of the spiraling, helical paths charged particles follow in an external magnetic field.

Subatomic particles behave the same way in the presence of each other's magnetic fields.

In addition, any torque exerted on the magnetic moment of the first particle, due to the magnetic field of the second particle, is equal and opposite the to the the torque on the second particle, due to the magnetic field of the first.

These equal and opposite torques may be considered as "central" between the two bodies.

The third symmetry of our model is -

The world is made up of equal parts matter and antimatter.
Matter spins to the left.   Antimatter spins to the right.

Our next "symmetry", concerns the microscopic and macroscopic 'regimes'.

In our model, the laws of physics are exactly the same in the subatomic realm and in the cosmic realm.

Time and space ~~behave~~ *are* exactly the same, everywhere, always.

The only difference between classical and quantum physics, is the nature of the _boundary conditions_ constraining the systems of interest.

The mantra of physics should continue to be;

Boundary conditions! Boundary conditions! Boundary conditions!

Hard to say and even harder to implement, but *very* important.

Our final "symmetry" is -

The gravitational and electromagnetic interactions behave exactly the same way
In both microscopic and macroscopic systems; the only difference between the two
(besides a giant scale factor), is the lack of a negative gravitational charge.

We whimsically call this 'broken' symmetry, e-Symmetry.

If this *is* a broken symmetry, then the culprit is, our 'newly discovered' phenomenon
called quantum mechanical electromagnetic induction.

The same phenomenon that breaks the symmetry of *classical* electrodynamics.

(We now take a brief pause for people to tape their brains back together)

:)

OK !

The universal reference frame:

In our model, the force between two bodies orbiting one another, normalized by the total energy of the system, is (14)

$$F/E_{TOT} = K*(c/R)^2\mu - K*(\mu v^2/R^2) - K*(l^2/\mu R^3) \tag{1}$$

We may write this symbolically as

$$F_{universal} = F_{central} + F_{coriolis} + F_{centrifugal} \tag{2}$$

These are all the "force terms". *However,* there is a contribution to the overall energy of the system, corresponding to the interaction of the spin/angular momentum (σ, l) of one object, with the 'magnetic force vector' of the other object.

$$E_{SPIN}/(E_{TOT}) = (l_1 \cdot B_2)/(m1+m2) + (l_2 \cdot B_1)/(m1+m2) \tag{3}$$

$$E_{universal} = E_{central} + E_{coriolis} + E_{centrifugal} + E_{spin} \tag{4}$$

These calculations must be performed in the universal reference frame as shown in Figure 1.

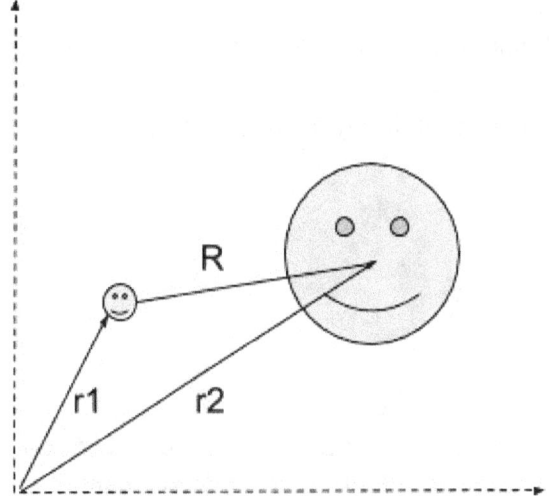

Figure 1: A cartesian coordinate system for the study of planetary motion. The coordinate system is in reference to the fixed background of space. The mutual force between the two bodies is a function of *all* relative velocities between the masses comprising the the two bodies.

Cosmology:

There are actually two more radial vectors necessary to completely describe the planetary motion depicted in Figure 1.

These are, of course, the radii of the orbiting bodies: r_A, r_B. Despite our previous admonitions, we *do* want to fix the coordinate systems for r_A and r_B at the center of the bodies A and B, respectively.

The relativistic mass of a celestial body includes a *considerable* contribution from its spin, and is dependent on the radius and rate of rotation of the body.

$$M = \Sigma_i \, m_i(v_i(\mathbf{r}_i)) \tag{5}$$

If we make the reasonable assumption that the angular velocities of the bodies are fixed during an interaction, we can make our first *approximation*, and perform the summation (integral) of equation (5) for the two individual bodies, before creating the reduced mass, μ, of equation (1).

We now have all the ingredients to do cosmology!

One chooses an 'isolated' system of interacting objects, and then applies the proper generalizations of equations (1) through (5).

All these interactions take place against the fixed 'backgrounds' of space and time.

"Sounds reasonable," you might say, "but how do you explain the gravitational redshift of light from a distant galaxy if space is a "fixed background" ?".

Gravitational redshift:

Since we no longer consider space to be a physical substance capable of expanding and contracting (it is now gone with the ether), we must assume that the gravitational redshift of light from a distant galaxy is due to the gravitational drag of the mass of the galaxy on the emitted photons.   The photons are now considered to have a gravitational charge proportional to their energy since we believe mass and energy are equivalent.

The universe is no longer flying apart, but we can still use the gravitational redshift of light to measure the distance of a faraway galaxy by calculating the gravitational work done on a photon by the mass of the emitting galaxy.   (A fun math project!)

Forces and potentials:

One of the recurring questions of physics is; which concept is more fundamental, the concept of force, or the concept of potential (energy).

Since the concept of force is more reflective of everyday interactions that human beings have with objects, and considering the fact the concept of potential is derived from the concept of force, we choose force to be the more fundamental concept.

Note, we purposely and repeatedly employ the term concept to emphasize that neither forces or potentials are fundamentally real.

Forces and potentials are useful abstractions invented by human beings to help them understand and describe the world.

In our model, the concept of potential *energy*, has been **banished**.

However, just like the concept of force, we retain the more formal concept of potential for utilitarian purposes.

Rather than speak of the interactions between particles in terms of the balance between the kinetic energy and the potential energy of particles, we characterize interactions solely in terms of the changes in the kinetic energy and *relativistic mass* of the particles.

In our model, even the work done in raising an object a distance, d, above the surface of the earth,

$$W = mgd \tag{6}$$

would actually be reflected and *formally* written as a change in the particle's relativistic mass-energy

$$m = m_{rest} c^2 + m_{rest} gd \tag{7}$$

However, equation (6) is the more appropriate and useful choice for day to day applications.

:)

The interaction between (two) bodies is completely determined by -- the mass of the two bodies, the relative velocity of the two bodies, and the relative separation of the two bodies.

Or, more generally, and more succinctly,

The interaction between bodies is completely determined by the relative relativistic mass-energy of the bodies, and the distance between the bodies.

This is illustrated for the classic, classical, two body system by equation (1).

It all sounds so very simple, except the relativistic mass-energy of (even) a two body system is a complicated function of the several and various, relative *and* absolute, velocities describing the motion of the individual bodies.

As usual (or, as it was in the 'good old days'), the physics is easy, but the math is hard!

However, it seems much of the mathematics, and many of the mathematical tools, generated during the investigations into the general theory of relativity, could 'easily' be "mapped over" or "reworked", and applied in the pursuit and development of equation (1).

synergy

On action and reaction:

The "action" is a well defined quantity and concept in terms of;

1) "Minimizing the action" (i.e. minimizing the time integral of the Lagrangian, T - V, of a system of particles), yielding a quantity with units of (energy) × (time); (E • t).

2) Planck's "quantum of action" with units of E • t , representing one unit (the smallest possible) of 'interaction' that may be exchanged between two particles;

'interaction' *defined* as the exchange of (a certain amount of) energy
during   (a certain amount of) time.

Everyone knows the units of the action, E • t, are equivalent to the units of angular momentum, kg m^2/s.

Particles interact by exchanging units of action, or units of angular momentum.

Moving angular momentum around takes *work*, and, of course, everyone wants to minimize their amount of work!   "Work smarter, not harder!"

Nature does not like to do work. Nature's motto is: Do as little work as possible !

This is the true anthropic principle.   :)

Newton's third law is often colloquially expressed as "for every action there is an equal and opposite reaction".

Of course, Newton's third law is actually a statement about forces, $F_2 = - F_1$, and the term "action" is not well defined. Action seems to be used metaphorically to say that everything that happens to 'object one' during an interaction also happens to 'object two' in a compensating manner, ensuring the conservation of energy, momentum, angular momentum, etc., during the interaction.

This concept of action and reaction is a more general and more inclusive description of the interaction between two bodies, compared to demanding a simple balancing of the central forces, $F_2 = - F_1$

In the universal model, Newton's third law will be broadened to include forces as well as "actions".

Just as we 'generalized' the classical Lorentz force (subsuming Newton's universal law of gravitation in the process!), we now 'universalize' Newton's three laws of motion.

1) A body will not experience a change in **momentum**, unless acted on by a force, **F**, generated by a second body (or several other bodies).

2) $\mathbf{F} = d\mathbf{p}/dt = d(m\mathbf{v})/dt = m\, d\mathbf{v}/dt + \mathbf{v}\, dm/dt$ (8)

3) The two bodies experience equal and opposite changes in their respective *actions* during the *interaction* (which we have defined as the *exchange of units of action.*)

$$F_1 \times d_1 \times t = -F_2 \times d_2 \times t \qquad (9)$$

and equal and opposite changes in angular momentum due to magnetic forces

$$\Delta \mathbf{L}_1 = -\Delta \mathbf{L}_2 \qquad (10)$$

We now have equivalent and compatible definitions for "the action" in both the classical and quantum formulations of our universal model.

Newton's universal law of gravitation is *replaced* by the more general, and *properly normalized*, equation (1), and all calculations are done in the universal inertial reference frame of Figure 1.

The concept of 'minimizing the action' is already the mathematical bridge between classical and quantum physics, and remains so in the universal model, although we expect all Lagrangians to eventually be formulated strictly in terms of ΔT !

Otherwise, we like to avoid the use of the concept of potential energy for formal conceptions and definitions, and save it for the the practical applications of calculating the energy states of real world systems, where various approximations are necessary.

So, in our model, we envision the following substitutions; for the total energy

$$E_{TOT} = T + V = const. \quad \Rightarrow \quad m = T + m_{rest} \qquad (11)$$

and for the Lagrangian

$$L = T - V \quad \Rightarrow \quad L = T - m_{rest} \qquad (12)$$

where T includes *all* kinetic energy; linear, rotational, etc.

Mass *is* resistance to acceleration. If a mass is successfully accelerated,
it tries to shed the extra energy, by emitting photons. Mass does not like to be accelerated!

Nature minimizes the action during an interaction, by keeping the 'relativistic'
masses of the particles involved, as small as possible.

Nature achieves change (as it strives for an ever elusive stability) by doing the
least amount of work possible, against all possible constraints, and within the
boundary conditions fixing the closed system and defining the
nature of the interaction.

To do the least amount of work, nature must keep acceleration to a minimum.

For any closed system, the sum of the individual energies divided by the total
energy of the system, is equal to one; and we may write, symbolically;

$$(T + m_{rest}) / m = 1 \qquad\qquad (13)$$

or more concretely,

$$((\Sigma_i T^i) + (\Sigma_i m^i_{rest})) / (\Sigma_i m^i) = 1 \qquad\qquad (14)$$

In this formulation, we see all conservative interactions must satisfy one constraint

$$\Sigma_i (\Delta T_i) = 0 \qquad\qquad (15)$$

During any interaction, or evolution of any closed dynamical system,
the sum of the changes in the kinetic energy of the particles involved,
must add to zero.

**In our model, _only relative changes matter_; relative changes in velocity
and kinetic energy.**

Newtonian versus Hamiltonian dynamics:

The chicken or the egg?

We have already concluded, the concept of force precedes (or at least, preceded) the concept of energy, and, as such, must be considered the more fundamental concept, and the starting point of all investigations.

In the case of Newtonian versus Hamiltonian dynamics, we know which came first.

The questions people still like to ask are;

"Is one a more primary picture of particle dynamics?"

"Is one picture more reflective of reality ? --
of how the world 'really' functions, and operates, and behaves, 'beneath it all' ?"

Our answer is both ironic and somewhat paradoxical.

The Hamiltonian approach is more in accord with the spirit of our model, although the model is derived from the concept of force (of course!), and depends on the concept of force for many or most real world applications.

Still, we chose the Hamiltonian approach, as the better picture of the general behavior of particles, and their interactions, and as a better description of the dynamical evolution of a system of interacting particles.

In our model, particles do not feel force, or exert forces on one another,
but rather, are constantly coupled, continually exchanging energy and momentum.

Let us say they are complementary.

On interaction:

In our model the fundamental particles are considered to be the physical incarnation, or manifestation, or *ensconcement*, of the fundamental unit of angular angular momentum; Planck's "quantum of action", h. This is the nature (or 'origin') of particle mass and spin.

This makes sense since intrinsic rotation, or spin, is translationally invariant, and apparently, compact, self-contained, and 'self-sustaining'. In addition, the intrinsic angular momentum of a particle (plus the precession of this angular momentum about the axis of the direction of motion) is the sole source, and origin, of inertial mass, relativistic mass, and gravitational mass.

We imagine the three fundamental particles as follows, as discussed in reference (14):

The photon is one 'free', massive, but inertialess, unit of angular momentum; L = h.

The neutrino is one 'bound', massive, unit of angular momentum *per unit space*; L = hbar/2.

The electron is one 'bound' unit of angular momentum per unit space *per unit mass*; L = hbar/2.

These three particles are the building blocks of matter in our model.

The leptons are the bricks, the 'virtual' photon is the mortar, and the 'real' photon is the wrecking ball!

The virtual photon is for 'internal bonding' only, in a 'closed' and 'conservative' system of interacting particles.

The real photon is for shedding unwanted energy and spin into the 'external' environment (thus, randomly wrecking things, or carrying the latest top forty tune to your car!).

Real photon *emission* is only necessary when our 'closed' system is perturbed by an external influence (e.g. in the electrification of hydrogen gas).

Either way, virtual or real,

There are three differences between the virtual photon and the real photon --

1) The virtual photon, spin 0, *transfers* energy and momentum *between* particles.
    The real photon, spin 1, *carries away* energy and momentum, *and* angular
    momentum (in units of h).

2) The virtual photon is imagined to be *a 'standing wave'*, of *varying frequency*,
    with the endpoints *secured* between two interacting, *moving* particles.
    The real photon is considered to be a free, 'traveling wave', of *fixed frequency*,
    emitted from an oscillating (and usually externally driven) source.

3) Virtual photons couple to the mass and spin of a particle.
    Real photons only couple to the spin of a particle;
                    i.e. the *magnetic moment*.

The proper wave equation for a real, *literally free*, (no 'scare quotes' necessary!)
photon is the Klein-Gordon equation. For simplicity, we assume travel along the x-axis,
and denote the real photon wave function by the capital greek letter Γ.

$$\partial^2\Gamma/\partial t^2 \;=\; \partial^2\Gamma/\partial x^2 \tag{16}$$

with solutions

$$E \;=\; +/- \;|\mathbf{p}| \tag{17}$$

$$\Gamma_{LEFT} \;=\; \varepsilon\exp(-ip\bullet x) \tag{18}$$

$$\Gamma_{RIGHT} \;=\; \varepsilon\exp(+ip\bullet x) \tag{19}$$

We imagine the the positive energy solutions to correspond to spin left solutions, or those
photons with negative (?) angular momentum, and the negative energy solutions to spin right
photons, or those with positive angular momentum.

These assignments are based on our assumption that the electron is considered matter
which spins to the left, while the positron is antimatter which spins to the right.

In our model, the photon is no longer its own antiparticle! There are now two distinct real
photons. One couples to 'matter', the other, 'antimatter'.

A real photon is one unit of angular momentum, or action; h. Photons spin to the left and the projection of this angular momentum is along the direction of travel.

Anti-photons spin to the right, and the projection of the the angular momentum is antiparallel to the direction of travel or propagation.

As for the mode of photon propagation, we originally imagined a 'flipping' of the photon polarization as simple harmonic oscillation, with the photon polarization vector oscillating sinusoidally solely along the direction of travel.

This flipping action was introduced to explain how different photons can have different energies and momenta even though for every photon the angular momentum and velocity are fixed quantities; L = h, v = c. Only the frequency varies.

Upon further reflection, we realized this picture of photon propagation can not be physical and is not correct. We now believe the 'flipping' of the photon polarization vector to be more akin to the 'spinor-ing' of the electron which we introduced in "On Action and Reaction".

The 'corpusculating' of the photon is shown in Figure 2.

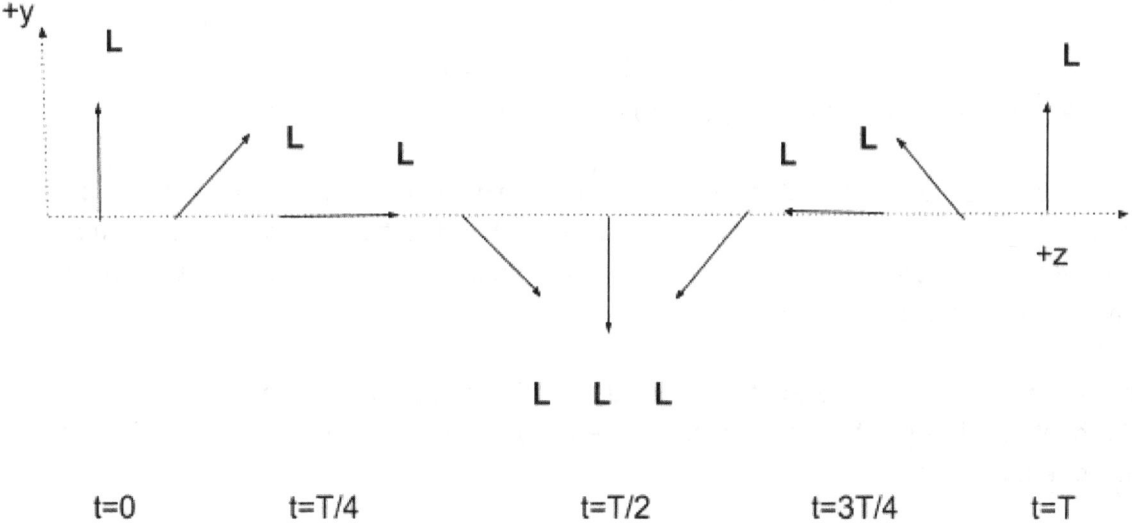

| t=0 | t=T/4 | t=T/2 | t=3T/4 | t=T |

Figure 2: A 'plane polarized' photon traveling in the z - direction. (Individual real photons always spin to the left or right. One may imagine the momentum vector L spinning about the z-axis.) The photon rolls, or 'corpusculates', along the direction of travel projecting the spin angular momentum sinusoidally along the direction of propagation while maintaining a constant polarization and angular momentum; L = h. We could also say that the photon 'spirals'.

The free photon satisfies the Klein-Gordon equation. In our interpretation, the positive and negative energy solutions correspond to positive and negative *helicity* solutions.

The free photon must also obey the Klein-Gordon probability current equation.

Again, instead of imagining positive and negative probability currents, we have two separate probability currents representing the two orthogonal, positive and negative, helicity solutions (i.e. left-handed and right-handed photons) of the KG equation.

We haven't quite worked out all the sign conventions describing the relations between the energy, angular momentum, helicity, and probability currents for photons and anti-photons, but the reader can easily see how the associations are to be made, at least in principle. (Homework!) We expect to present a more quantitative discussion of helicity and anti-particles in our next book "On Rotation".

A real, free photon always spins to the left or the right. Therefore, the wave function of a real, free photon must be a linear combination of the two orthogonal, transverse solutions of the KG equation. Does this suggest the wrong 'eigenbasis'; or perhaps that the Coulomb gauge is not the best choice for describing free photons?

To further investigate the behavior of the photon and the nature of photon interactions in our new model, let's look again the theory of photon "corpusculation" as depicted in Figure 2.

At time, t=0, we happen to catch the photon in the middle of a 'helicity flip'. At this moment, the photon angular momentum is entirely *transverse* to the direction of propagation. If the photon were to meet an electron at this particular moment, it could not interact because its angular momentum is 'unavailable'.

At time, t=T/4, the photon angular momentum is fully projected along the direction of travel, and the photon is in a position to transfer all its energy and momentum to an available and suitably oriented electron.

In our model, photons are self-coupling gauge bosons.
A photon and an anti-photon can annihilate producing a virtual photon;

(spin +1) plus (spin -1) = spin 0

In our theory, two properly oriented "two component" real photons can combine to form one "four component" virtual photon. It seems this particular virtual photon would have no longitudinal component . . . Just musing at the moment !

More generally, of course, a virtual photon connecting two electron has four components. This photon appears as a 'double spiral', like a double spiral staircase, with two components carrying energy and momentum in the **+r** direction, the other two in the **-r** direction.

Quantum mechanical electromagnetic induction:

In our model, not only do elementary particles really spin; they essentially *are* spin.

When it was first suggested that the electron might have a spin, it was decided that this was merely a 'quantum label' and could not be a *proper* classical spin, because the electromagnetic force was not sufficient to keep the electron from flying apart. This idea has not served physics well.

We now have the Magic of Quantum Mechanical Electromagnetic Induction to endow the electron with mass and electric charge, as well as stability and internal coherence.

Electromagnetic induction was the last classical electromagnetic phenomena to be discovered. It is not at all obvious, and cannot be derived from, or proved from, the other mathematical laws of electricity and magnetism.

Electromagnetic induction is the only classical phenomena that has not been observed in, or contemplated for, the quantum realm.

From arguments of beauty, symmetry, etc., we propose that electromagnetic induction is one of the fundamental laws of nature and must have a quantum mechanical counterpart.

Thus, quantum mechanical electromagnetic induction *just is a thing*, needing no further explanation or explication, and it gives rise to the electron and the electric charge.

Our speculations on quantum mechanical electromagnetic induction and the nature of the electron have been presented in several previous publications (references 3, 6, 7, 14).

We can summarize here by simply saying;

The electron is a puffed up neutrino.

The Dirac equation:

As previously stated in this paper and elsewhere, in our model all interaction is determined solely by the relative motion of massive bodies and can be completely characterized by the conservation of kinetic energy as demonstrated by equation (15).

In physics, we are not interested in the (essentially arbitrary) absolute 'total energy' of a particle or a system of particles, but rather in the *change* in energy over a prescribed period of time of interaction. In our model, any change in energy is a change in *kinetic energy*.

In this spirit, we will suggest a method to try and remove the rest mass from the Dirac Equation/Lagrangian in a 'rigorous' way.

The Dirac equation for a 'free electron' is

$$H \, \psi = (\, \alpha \bullet p + \beta \, m_0 \,) \, \psi \qquad (20)$$

where the Hamiltonian H, is the total energy of the electron, and m_0 is the rest mass. The Hamiltonian operator is

$$H = i \, \partial / \partial t \qquad (21)$$

Let's define the kinetic energy operator (T = m - m_0) to be

$$T = H - \beta \, m_0 \qquad (22)$$

$$T = \alpha \bullet p \qquad (23)$$

In our model, the lagrangian is $L = \Delta T$, and

$$\Delta T = H \, | \,_{t=t2} - \, H \, | \,_{t=t1} \qquad (24)$$

Admittedly, this discussion is still a bit muddled … maybe we'll have more later.

The real point we'd like to make about the Dirac equation today is that the primary operator and eigenvectors characterizing the Dirac spinor should be the observable $\sigma \bullet \mathbf{p}$ with the helicity eigenvectors +/- ½.

In this approach, just as in our model of the photon, the positive and negative 'energy' solutions of the Dirac equation are now interpreted as the (positive) energy solutions for the positive and negative eigenstates of the helicity equation (i.e. positrons and electrons).

To remind the reader of our definition of electrons and positrons, and the corresponding concept of electron propagation as 'spinoring', we reproduce here the diagram first introduced in "On Action and Reaction" as Figure 3.

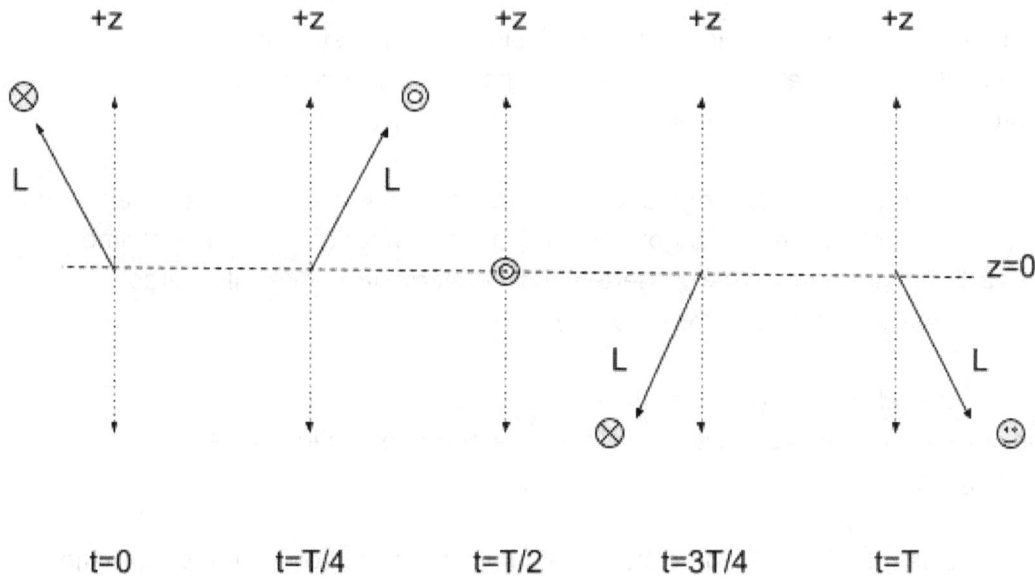

Figure 3: At rest with an electron traveling the the z-direction. The spin angular momentum vector 'precesses' about the direction of motion, tracing out a closed, three dimensional figure eight. The x symbol represents motion into the page. The dot symbol represents motion out of the page. At time T/2, we see the the angular momentum is *perpendicular* to the direction of travel. (This is when the electron engages in 'virtual' interactions.)

To represent a positron, simply swap the x symbols and the dot symbols.

In Figure 3, we can begin to see inklings of the uncertainty principle. Ideally, at time, t=0, the energy of the electron can be measured exactly. However, we cannot make instantaneous measurements as measurements always take some finite amount of time.

At the other extreme, at time, t=T/2, the electron is unable to interact with our apparatus at all!

In between, the fraction of energy available for interaction and measurement varies sinusoidally.

Und so weiter.

Wave particle duality:

When you want to study the 'wave like nature' of the photon or the electron, you must employ the particle wave function 'in real time' in order to include the contributions from all local and global phase factors influencing an interaction.

The particle phase, previously dismissed as a curiosity, actually determines the timing, place, and nature (i.e. 'real' versus 'virtual') of a particle's interaction with neighboring matter.

When studying the wave like nature of particle interactions, we must use the real part of the imaginary wave function, yielding trigonometric functions that describe the real time behavior of the particles, predicting when, where, and in what manner they will interact with our detector (the proverbial photographic plate).

When you want to study the particle like nature of the photon or the electron, you employ the square of the wave function, yielding average expectation values for position, velocity, etc.

Let's consider the single slit diffraction experiment using our model of photon propagation and interaction as depicted in Figure 2. If the photon arrives at the photographic plate at time, t=0, it cannot darken the plate. The photon energy and momentum are unavailable for 'real' interaction at this instance. If the photon arrives at time, t=T/4, it will slam the photographic plate full force!

The root of the wave particle duality 'problem', lies in the declaration, or decision (i.e. the *interpretation*), that quantum spin is some crazy, special feature of microscopic particles; that it has no counterpart in the macroscopic world; does not, and *could* not, correspond, in *any* possible, conceivable, or imaginable way, to our average, ordinary, everyday, common sense ideas, and notions, of rotation.

In our model, particles are real, spin is real, and particles *really* spin.

Occam's razor to the rescue !

Quantization:

There are no quantum variables that we do not also observe at the macroscopic level. Quantum mechanics deals with mass, spin, charge, magnetic moments, and the conservation of energy and momentum; all phenomena which have a classical counterpart.

In our model, there are no flavors, no colors, no fractional charges, no charged fields, no mass inducing fields (no fields at all!), no ghostly evolution of particle states, no vacuum, no "spin" that doesn't really spin, etc., etc.

As Werner Heisenberg has pointed out, if there were quantum variables that did not correspond to a familiar macroscopic quantity, we would not be able to recognize them.

All the classical concepts and quantities that we use to describe the macroscopic world are quantized at the microscopic level, including mass, the electromotive force, and the magnetic moment of a particle. Furthermore, even acceleration seems to be quantized as evidenced by the electronic transitions in the hydrogen atom, for example.

The only difference between macroscopic, classical physics, and microscopic, quantum physics, is the nature of, and satisfaction of, the boundary conditions of the physical system.

We are allowed to speak of the 'quantization of space'; e.g. in characterizing the electron orbits of the hydrogen atom, but it should be clear this quantization is a property of the relation between the particles, and not a feature of space itself.

Quantized space is *not* chunky.

Ultimately, quantization boils down to the satisfaction of boundary conditions.

Virtual photons can only attach to/interact with an electron when the electron has an effective helicity of zero (during its polarization flip) because virtual photons are spin zero.

This fixes the 'frequency' of the virtual photon coupling two interacting electrons, since virtual interactions, or the exchange of energy and momentum, can only occur when both electrons have helicity ~= 0.

When an electron reaches 'peak helicity', it is able to interact with (i.e. absorb and emit) real photons the most readily and the most efficiently.

The special theory of relativity:

The special theory of relativity is concerned with point-like, spacetime 'events', and whether these may be considered simultaneous in various reference frames and in what regard.

In the *universal theory of relativity*, our concern is with the *interaction <u>between</u> particles <u>over some finite amount of time</u>*, and *interactions* are simultaneous in any reference frame.

You cannot do physics with one particle, nor at a single point in space and time.
For every action there is an equal and opposite reaction. In any reference frame.

All interactions occur at the speed of light regardless of the choice of reference frame. *Nature does not know about reference frames.* 'Nature' only 'cares' about the relative positions and velocities of interacting bodies.

The (background of) space in which these interactions occur is fixed, immutable, absolute, and empty; uniform, smooth, flat, and three dimensional; homogeneous, isotropic, and undifferentiated.

Time and space are *absolute*. Only *coordinate systems* are relative.

'Observers' in different inertial reference frames measure the relevant quantities of an interaction using different position *and* time variables; however, they describe the same world, in the same way, with the same results.

The crucial point here is that the different observers measure different *time intervals* **and** different *space intervals* for any given event. This fact takes a bit of the shine off of any mind boggling interpretation of time dilation. If different observers, in different inertial reference frames, measured *different time intervals*, and *yet **identical space intervals***, this would be a truly mind boggling break with Galilean relativity.

Most (of my favorite!) authors seem compelled to convince the reader (and themselves?) that time dilation is not merely appearance, but something real, with real, physical, and measurable effects. They all chose to cite the (only?) example of long lived muons resulting from high energy cosmic ray collisions in the earth's upper atmosphere.

In this example, high energy (v/c ~= 0.98) muons are believed to travel longer distances than usual in *our* reference frame because they have a longer lifetime (due to their velocity) in *their* reference frame!

This is clearly nonsense! In the moun reference frame, the muon is at rest, and thus has the usual 'rest mass', or 'proper', lifetime. The fact is the muon is traveling at v/c = 0.98 in *our* reference frame

Time dilation *is* mere appearance, and cannot account for long lived, high energy muons.

In our model, there is no 'phase space' for the decay of high energy muons because they are too massive (14).

The muon must make multiple collisions before reaching 'decay velocity'.

We remind the reader of our theory of muon decay by including the diagram first introduced in "On Parity and Isospin" as Figure 4.

In our model, the muon decays into an electron by 'shedding' excess energy and momentum in the form of its neutrino. The resulting 'propagator' **_must be_** a massless 'virtual lepton'.

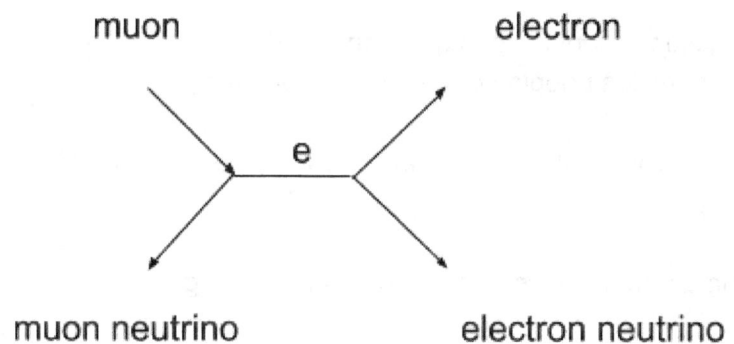

FIGURE 4: Muon decay. The propagator is actually a generic virtual lepton, e/mu/tau.

The energy, momentum, and spin shed by the moun ensures that the 'virtual lepton propagator' has a spin = 0, and is *massless*. This decay process is thus kinematically prohibited for ultra-high energy muons !

The standard model:

In "A Quantum Mechanical Theory of Everything" we said:

The two great pillars of twentieth century physics, general relativity and quantum mechanics, are incompatible and irreconcilable because they are both incorrect.

Of course, we were just having fun with this shopworn cliche.

In our model, general relativity *is* incorrect and we see nothing to recommend it.

On the other hand, we would *not* say quantum mechanics is incorrect except in this particular context. Here, 'quantum mechanics' is taken to mean 'the standard model' and/or quantum field theory, *plus* every possible spooky speculation ever spouted by anyone.

In our view, quantum mechanics is bras, kets, state vectors, wave equations, operators, matrix elements, and expectation values.

Quantum mechanics is *not* wave function collapse, spooky action at a distance, particles appearing two places at once, particles popping out of the vacuum, etc.

In our model, we *reject* quantum field theory while retaining all the rational bits of quantum electrodynamics.

The result is 'simple' relativistic quantum mechanics. The theory of everything.

I believe we have sufficiently proven that gauge theory is incorrect and that gauge symmetry is not a feature of the world (14), that the quantum vacuum is not a thing (6), and that fields of any description are *not* real, but only useful fictions.

Regardless, gauge theory and the magical, miraculous vacuum can no longer come to the rescue of QFT (e.g. renormalization, etc.), so it is a lost cause.

Conclusion:

One *cannot* physically or metaphysically interpret mathematical equations.

The mathematical equations of physics, are *already* (*or should be*) a self-contained and coherent interpretation of all currently known physical data; a codification of the well confirmed, and well understood, observations of how physical systems behave.

Math, as the language of physics, is *already* an interpretation; an expression of knowledge, not easily, or usefully codified, in spoken or written language alone.

The mathematical equations of physics are for making *physical predictions* that can be *tested* by clever and well designed experiments.

They may not be used to make metaphysical assumptions about the world.

Of course, one may fiddle about and accidently discover an equation, then find the equation may be construed to explain several previously known and puzzling facts; the Dirac equation being one of the most famous examples.

The discovery of the Dirac equation was a truly great and momentous invention!

The invention of the Dirac equation was a truly great and momentous discovery!

Truly!

Then, the *interpretations* began, giving us an infinite sea of antiparticles traveling backwards in time (obviously!), just beyond our reach due to some great rift, or potential barrier, between the negative energy quantum vacuum and the positive energy quantum vacuum of our real world.

Better to "make no hypothesis".

the sinister universe:

iconographic
apples   !

(*yawn*)

no tumbling dice

no cats,
ineffably
entombed

to die

(can you even *put* a cat in a box?)

no black box devices

no malicious g-d
no master plan.

only masses,
matter

in motion,
thru space

in time

hand in hand

Books by Greg Feild: The SInister Universe Series

the pentateuch

1. "A quantum mechanical theory of gravitational interactions"
   CreateSpace Independent Publishing, 8/29/2016

2. "Observations on the quantum mechanical nature of gravity"
   CreateSpace Independent Publishing, 10/8/2016

3. "On gravitation and electric charge"
   CreateSpace Independent Publishing, 10/29/2016

4. "On spin, mass, and charge"
   CreateSpace Independent Publishing, 11/29/2016

5. "On angular momentum, acceleration, and absolute motion"
   CreateSpace Independent Publishing, 1/1/2017

the exegeses

6. "The Sinister Universe"
   CreateSpace Independent Publishing, 3/1/2017

7. "On Parity and Isospin"
   CreateSpace Independent Publishing, 4/11/2017

8. "Reflections on the Sinister Universe"
   CreateSpace Independent Publishing, 5/12/2017

the hermeneutics

9. "On Current Physics"
   CreateSpace Independent Publishing, 6/11/2017

10. "A Critical Examination of Classical and Quantum Mechanical Waves"
    CreateSpace Independent Publishing, 6/18/2017

the gospels   :)

11. "On wave particle duality and the quantum of action"
    CreateSpace Independent Publishing, 7/6/2017

12. "On matter, mass, and motion"
    CreateSpace Independent Publishing, 9/14/2017

13. "On action and reaction"
    CreateSpace Independent Publishing, 9/24/2017

14. "A quantum mechanical theory of everything"
    CreateSpace Independent Publishing, 11/5/2017

the compilations

"The Universal Model of Our Sinister Universe: The First Ten Books"
CreateSpace Independent Publishing, 7/2/2017

"The Canons of the Sinister Universe:
 The Last Four Books on the Universal Model of Our World"
CreateSpace Independent Publishing, 11/5/2017

Alice's Evidence:

"No, no!", said the Queen.
"Sentence first - verdict afterwards."

"Stuff and nonsense!", said Alice loudly.
"The idea of having the sentence first!"

    ↺   ↻   ↺   ↻

"Who cares for you?", said Alice.
"You're nothing but a pack of cards!"

              -- Lewis Carroll
                Alice's Adventures in Wonderland

White Rabbit:

When logic and proportion
Have fallen sloppy dead
And the White Knight is talking backwards
And the Red Queen's off with her head ...

Remember - - - - what the Dormouse said

Feed your head

              -- Grace Slick

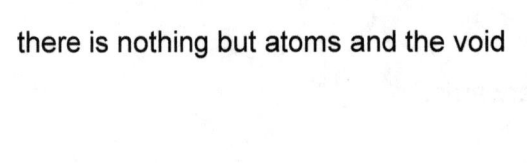

www.ingramcontent.com/pod-product-compliance
Lightning Source LLC
Chambersburg PA
CBHW080629220526
45467CB00011B/3424